m&s

minutes&seconds
theScientists

patrickaievoli

Zea Books
Lincoln, Nebraska
2018

Minutes & Seconds: The Scientists
Patrick Aievoli

Zea Books, Lincoln, Nebraska, 2018

ISBN 978-1-60962-130-8

doi 10.13014/K2FT8J8M

Zea Books are published by the University of Nebraska–Lincoln Libraries.

Electronic (pdf) edition available online at http://digitalcommons.unl.edu/zeabook/

Print edition available from http://www.lulu.com/spotlight/unlib

UNIVERSITY OF
Nebraska
Lincoln®

Dedication

This work is dedicated to my mother and father.

They both knew they had a bit of a "strange one"
on their hands at an early age.

But, nevertheless their support and guidance
never waivered.

Thank you Josephine and Patsy Aievoli.
A perfect team of love and caring.
As I said before "he pedaled and she steered".

"Love is forever."

Foreword

Minutes & Seconds:

In creating Minutes & Seconds, Aievoli has assembled an interesting compilation of scientists and their respective inventions or contributions that have not only changed the world as we know it, but have stretched our intellect and imaginations.

With careful thought selection and design embeds, the artist and author treats us to both a history lesson and visual storytelling that challenges us to consider our past innovations, the works of genius, that have brought us to where we are today -technologically speaking. By so doing, it's not hard to see the author's underlying commentary about where technology has taken us, with both positive and negative effects.

With equal parts homage to his scientific heroes, and healthy skepticism of modern day opportunists, Aievoli makes keen observations about the lasting societal impact of past inventions and the intent of the men and women behind them. Conversely, he also raises subtle questions about how we have coopted technology for not so quite as noble intentions; how profiteers have traded scientific and intellectual advancement for thought manipulation, control of information, and forced dependency.

Who are the best minds we have today? What are they inventing to help humanity progress? Take a broad look at the overall intellectual landscape and ask yourself has technology (and our dependency on our gadgets) made us smarter, or lazy?

Because we so readily use technology to fill in for lack of knowledge, or to entertain, distract, manage online identities, do we even leave time for ourselves to think? Or, have we lost the ability to create, innovate, invent? Where are the tinkerers, explorers, or scientists of our day? If we lose those minds, those talents, how will we continue to compete globally?

These are interesting times in which we live, and interesting concepts to consider in the face of our ever-changing technological designs and their possible effects on human thought and productivity. This book is a beautiful reminder to be ever vigilant, ever aware of where we are letting technology take us!

Jennifer Cusumano, M.A., M.S., Ed.D.

Preface

To a designer the concept of three (3) is holy in its own right. In a successful visual design there are always three components – usually dominant, sub-dominant and focal point. In a successful painting it is called triadic harmony, always keep the viewer moving about the page from component to component, while using that as motion to keep them from diverting off. This is true in most all other forms of design; architecture, product and industrial.

The three areas I am also referencing are in the concept stage of creation; the theory, the application of that theory and the inspiration of the original concept - the "what if?" moment. That is really the core of this series "minutes & seconds" – the moment when the pattern reveals itself.

Having been an educator for over 30 years and a designer for over 40 the recognizing of patterns has always been my main "go to" element in my skill set. The engine that has pushed my career along has always been pure curiosity – the "what if" moment. Today that moment – I believe – is gone.

I love technology – always have – since early Flash Gordon and Star Trek, Star Wars whatever vehicle how we progress has always intrigued me. The closer the fiction was to real theories the better. I was always fascinated at how diverse theories were linked together to create anew. Simply put – the question of "what if"?

Today in my opinion technology has killed off that moment in society. Today we have the answers so quickly we seem not to have the time to question. "Professor Google" is right there with the answer and we move on to SnapChat with our "friends" and show our newest filtered image.

The use of these digital Libraries of Alexandria has been short sheeted to "I know the answer I don't want to remember it nor question it…" Kimye has a new selfie up and she/they want to be my "friend".

It is like the libraries have just taken the old card catalogues and dumped them into a big pile with no referencing or linkage. Today we wander through these heaps grabbing a bit of recorded knowledge and then feel like our task has been successfully achieved and we move on.

This series – this book – has been structured not to give a history lesson nor a technology lesson but a simple re-invigorating of the question – "what if?" In Italian it is summed up as "cosa succede se?" or "Io non lo so". Later on in college I came to understand this is called the Socratic method-asking questions.

When Newton developed his theories they were based off of his prior knowledge base but probably using the question of "what if?" then Faraday used those theories and postulated his own "what if?" and on to Maxwell and Einstein and well – they told two friends and so on and so on….

Why entitle the series "minutes & seconds"? As Einstein supposedly said, *"we invented time so that everything doesn't happen at once"*.

Time is a very confusing factor when combined with events. In the world of interaction design time and events create the "reality". Obviously in life this is also true. Life is a sequence - and we go from one event to another. What I am fascinated by is the patterns developed by that sequence, the "what if?" moments that occur once those patterns are recognized.

The illustrations I created for this book are of individuals that in my opinion represent the points on that line of events. In some cases they are tangential, and in some cases sequential.

And like those events that formed their successor these illustrations are created as sort of a series of "digital collages" using the invention of the Knoll Brothers. In the heritage of Schwitters, Heartfield and Rauschenberg, I don't claim ownership of the original images used in these collages, I don't claim ownership of the articles that accompany these illustrations neither, nor the concept of a "collage" nor the concept of a book, but how I put them together is the key part. For that is the essence of design and I consider myself a designer not just an artist. To me art is just one of the tools a designer uses to solve a problem.

Engelbart didn't invent electricity nor did he invent the mechanics used, but he did put those concepts together to create a solution - the computer mouse. And Adele Goldberg and her team at PARC used that solution to develop the GUI (Graphical User Interface) that we all use today - everyday.

Listed below are some of the patterns I recognized to solve the problem of - "how we got to where we are in regards to the current use (and abuse) of technology in our lives." The quotations below are taken from a diverse range of my research for this book and I have hopefully strung them together to support my own "what if?" moment.

In his Principia Mathematics, Newton gave the foundations of classical mathematics and gravity.

Maxwell's electromagnetic theories, and the resulting equations, were consequently the greatest advance in scientific knowledge since Newton's Principia.

Then the story of two brilliant nineteenth-century scientists who discovered the electromagnetic field, laying the groundwork for the amazing technological and theoretical breakthroughs of the twentieth century.

With Herschel, Babbage worked on the electrodynamics of Arago's rotations, publishing in 1825. Their explanations were only transitional, being picked up and broadened by Michael Faraday.

Ada [Lovelace] was fascinated by Babbage's ideas. Known as the father of the computer, he invented the difference engine, which was meant to perform mathematical calculations.

Maxwell had originated the idea of representing physical phenomena with a field theory governed by mathematical equations, and in tribute to this Einstein later said: This change in conception of reality is the most profound and the most fruitful that physics has experienced since the time of Newton.

And so on, and so on...

Patterns will always amaze me but they do propose a larger force at work. One day we will all find out what that force is and what the plan was that they had in mind. And I do firmly believe there was or is a plan because as Einstein also supposedly said *"I, at any rate, am convinced that He does not throw dice."*

Patrick Aievoli - 2018

I

Sir Isaac Newton
(1643 - 1727)

Isaac Newton (January 4, 1643 to March 31, 1727) was a physicist and mathematician who developed the principles of modern physics, including the laws of motion, and is credited as one of the great minds of the 17th century Scientific Revolution. In 1687, he published his most acclaimed work, Philosophiae Naturalis Principia Mathematica (Mathematical Principles of Natural Philosophy), which has been called the single most influential book on physics. In 1705, he was knighted by Queen Anne of England, making him Sir Isaac Newton.

Isaac Newton's Discoveries

Newton made discoveries in optics, motion and mathematics. Newton theorized that white light was a composite of all colors of the spectrum, and that light was composed of particles. His momentous book on physics, Principia, contains information on nearly all of the essential concepts of physics except energy, ultimately helping him to explain the laws of motion and the theory of gravity. Along with mathematician Gottfried Wilhelm von Leibniz, Newton is credited for developing essential theories of calculus.

Read More at:

https://www.biography.com/people/isaac-newton-9422656

Isaac of all trades. English scientist and mathematician Isaac Newton is most famous for his law of gravitation, and was instrumental in the scientific revolution of the 17th century. Above: A photo of Newton Investigating Light, a portrayal of Isaac Newton created by J A Houston, circa 1879. (Photo by UniversalImagesGroup.)

(Photo: Universal History Archive/Getty Images)

1687 57

Isaac Newton

ANALYSIS

SERIES, FLUXIONES,

"To myself I am only a child playing on the seashore, while vast oceans of truth lie undiscovered before me."

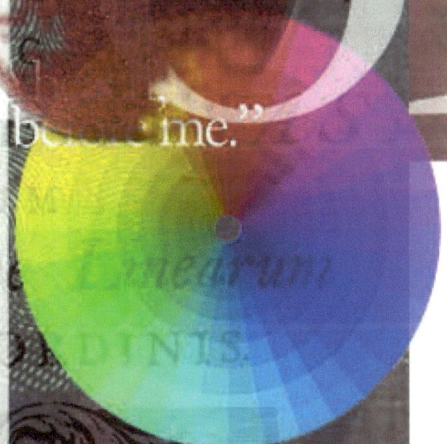

tanding on the shoulders giants

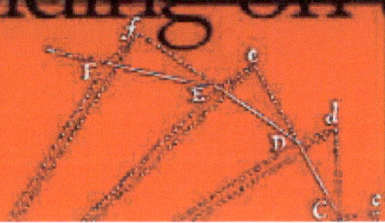

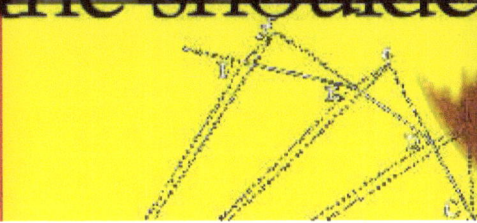

© 2018

2

Michael Faraday
(1791-1867)

English physicist and chemist whose many experiments contributed greatly to the understanding of electromagnetism.

Faraday, who became one of the greatest scientists of the 19th century, began his career as a chemist. He wrote a manual of practical chemistry that reveals his mastery of the technical aspects of his art, discovered a number of new organic compounds, among them benzene, and was the first to liquefy a "permanent" gas (i.e., one that was believed to be incapable of liquefaction). His major contribution, however, was in the field of electricity and magnetism. He was the first to produce an electric current from a magnetic field, invented the first electric motor and dynamo, demonstrated the relation between electricity and chemical bonding, discovered the effect of magnetism on light, and discovered and named diamagnetism, the peculiar behaviour of certain substances in strong magnetic fields. He provided the experimental, and a good deal of the theoretical, foundation upon which James Clerk Maxwell erected classical electromagnetic field theory.

Read More at:

https://www.britannica.com/biography/Michael-Faraday

$$EMF = -\frac{d\Phi}{dt}$$

$$\int_S \nabla \times \mathbf{E} \cdot \mathbf{dS} = -\frac{d}{dt}\int_S \mathbf{B}(t) \cdot \mathbf{dS} = \int_S \frac{-d\mathbf{B}(t)}{dt} \cdot \mathbf{dS}$$

$$\Rightarrow \boxed{\nabla \times \mathbf{E} = \frac{-\partial \mathbf{B}(t)}{\partial t}}$$

mifche

ulation

oder

practifche der fichern
chemifcher Arbeiten
experimente,

von

Faraday,

er Royal Institution of Great Britain
ler gelehrten Gefellfchaften.

Englifchen.

K:K:ART.
AKADEMIE

7.12.1812

faraday

$$\chi = \frac{\mu_0 + \nu\mu}{B}$$

400   500   600   700
Wavelength (nanometers)

3

Charles Babbage
(1791-1871)

Charles Babbage (1791-1871), computer pioneer, designed two classes of engine, Difference Engines, and Analytical Engines. Difference engines are so called because of the mathematical principle on which they are based, namely, the method of finite differences. The beauty of the method is that it uses only arithmetical addition and removes the need for multiplication and division which are more difficult to implement mechanically.

Difference engines are strictly calculators. They crunch numbers the only way they know how - by repeated addition according to the method of finite differences. They cannot be used for general arithmetical calculation. The Analytical Engine is much more than a calculator and marks the progression from the mechanized arithmetic of calculation to fully-fledged general-purpose computation. There were at least three designs at different stages of the evolution of his ideas. So it is strictly correct to refer to the Analytical Engines in the plural.

Read More at:

http://www.computerhistory.org/babbage/engines/

PORTION OF BABBAGE'S DIFFERENCE ENGINE.

4

Ada Lovelace
(1815-1852)

Ada Lovelace (1815-1852) was born Augusta Ada Byron, the only legitimate child of Annabella Milbanke and the poet Lord Byron. Her mother, Lady Byron, had mathematical training (Byron called her his 'Princess of Parallelograms') and insisted that Ada, who was tutored privately, study mathematics too - an unusual education for a woman.

Ada met Babbage at a party in 1833 when she was seventeen and was entranced when Babbage demonstrated the small working section of the Engine to her. She intermitted her mathematical studies for marriage and motherhood but resumed when domestic duties allowed. In 1843 she published a translation from the French of an article on the Analytical Engine by an Italian engineer, Luigi Menabrea, to which Ada added extensive notes of her own. The Notes included the first published description of a stepwise sequence of operations for solving certain mathematical problems and Ada is often referred to as 'the first programmer'. The collaboration with Babbage was close and biographers debate the extent and originality of Ada's contribution.

Read More at:
http://www.computerhistory.org/babbage/adalovelace/

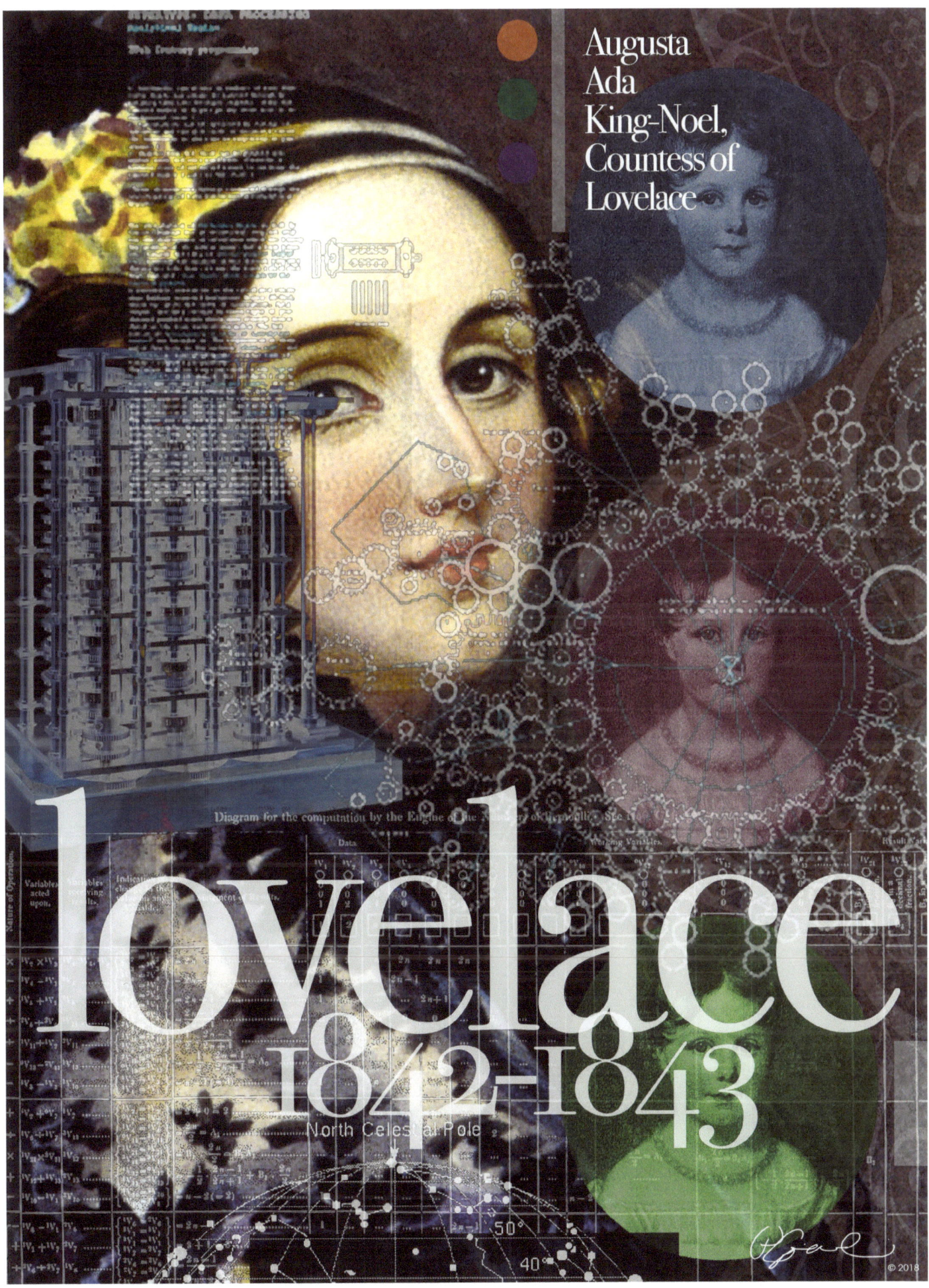

Augusta
Ada
King-Noel,
Countess of
Lovelace

lovelace
1842-1843

North Celestial Pole

© 2018

5

James Clerk Maxwell
(1831-1879)

James Clerk Maxwell was a Scottish physicist, best known for developing electromagnetic theory. As a child he was interested in colour and shapes, which attracted him to study mathematics and natural philosophy. He wrote his first scientific paper at the age of 14, which was presented to the Royal Society of Edinburgh on his behalf as he was too young to attend. He studied at the universities of Edinburgh and Cambridge, supplementing this by carrying out his own experiments and studies during summer holidays on his family's Scottish estate.

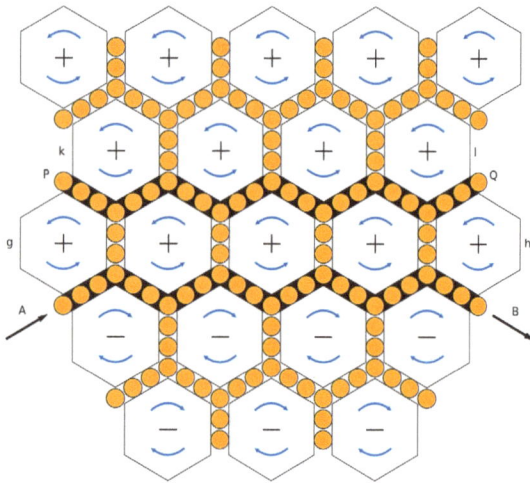

Maxwell became Professor of Natural Philosophy at Marischal College, Aberdeen in 1856. While at Aberdeen he studied the dynamics of Saturn's rings, proposing that they were made up of many small particles rather than being solid. This was confirmed experimentally several decades later.

Read More at:

http://www.sciencemuseum.org.uk/onlinestuff/People/James Clerk Maxwell 183179.aspx

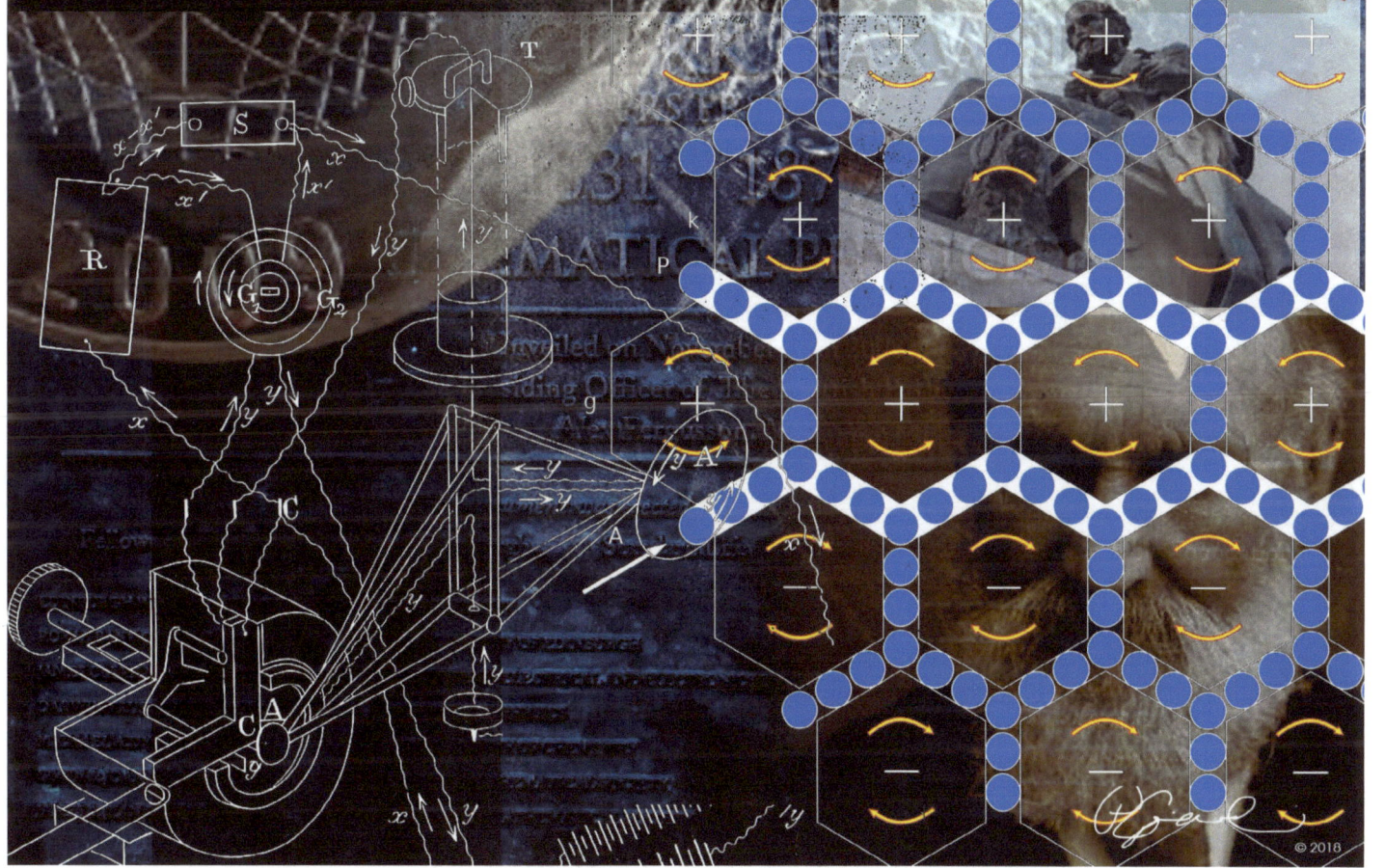

6

Albert Einstein
(1879-1955)

Albert Einstein was born at Ulm, in Württemberg, Germany, on March 14, 1879. Six weeks later the family moved to Munich, where he later on began his schooling at the Luitpold Gymnasium. Later, they moved to Italy and Albert continued his education at Aarau, Switzerland and in 1896 he entered the Swiss Federal Polytechnic School in Zurich to be trained as a teacher in physics and mathematics. In 1901, the year he gained his diploma, he acquired Swiss citizenship and, as he was unable to find a teaching post, he accepted a position as technical assistant in the Swiss Patent Office. In 1905 he obtained his doctor's degree.

During his stay at the Patent Office, and in his spare time, he produced much of his remarkable work and in 1908 he was appointed Privatdozent in Berne. In 1909 he became Professor Extraordinary at Zurich, in 1911 Professor of Theoretical Physics at Prague, returning to Zurich in the following year to fill a similar post. In 1914 he was appointed Director of the Kaiser Wilhelm Physical Institute and Professor in the University of Berlin. He became a German citizen in 1914 and remained in Berlin until 1933 when he renounced his citizenship for political reasons and emigrated to America to take the position of Professor of Theoretical Physics at Princeton*. He became a United States citizen in 1940 and retired from his post in 1945.

Read More at:

https://www.nobelprize.org/nobel_prizes/physics/laureates/1921/einstein-bio.html

$$E = mc^2$$

Energy is equal to mass multiplied by the speed of light squared

Thomas Alva Edison
(1847-1931)

In 1869, at 22 years old, Edison moved to New York City and developed his first invention, an improved stock ticker called the Universal Stock Printer, which synchronized several stock tickers' transactions. The Gold and Stock Telegraph Company was so impressed, they paid him $40,000 for the rights. With this success, he quit his work as a telegrapher to devote himself full-time to inventing.

By the early 1870s, Thomas Edison had acquired a reputation as a first-rate inventor. In 1870, he set up his first small laboratory and manufacturing facility in Newark, New Jersey,

Credit: General Eectric

and employed several machinists. As an independent entrepreneur, Edison formed numerous partnerships and developed products for the highest bidder. Often that was Western Union Telegraph Company, the industry leader, but just as often, it was one of Western Union's rivals. In one such instance, Edison devised for Western Union the quadruplex telegraph, capable of transmitting two signals in two different directions on the same wire, but railroad tycoon Jay Gould snatched the invention from Western Union, paying Edison more than $100,000 in cash, bonds and stock, and generating years of litigation.

In 1876, Edison moved his expanding operations to Menlo Park, New Jersey, and built an independent industrial research facility incorporating machine shops and laboratories. That same year, Western Union encouraged him to develop a communication device to compete with Alexander Graham Bell's telephone. He never did. However, in December of 1877, Edison developed a method for recording sound: the phonograph. Though not commercially viable for another decade, the invention brought him worldwide fame.

Thomas Edison, the Light Bulb and Electricity
While Thomas Edison was not the inventor of the first light bulb, he came up with the technology that helped bring it to the masses. Edison was driven to perfect a commercially practical, efficient incandescent light bulb following English inventor Humphry Davy's invention of the first early electric arc lamp in the early 1800s. Over the decades following Davy's creation, scientists such as Warren de la Rue, Joseph Wilson Swan, Henry Woodward and Mathew Evans had worked to perfect electric light bulbs or tubes using a vacuum but were unsuccessful in their attempts.

After buying Woodward and Evans' patent and making improvements in his design, Edison was granted a patent for his own improved light bulb in 1879. He began to manufacture and market it for widespread use. In January 1880, Edison set out to develop a company that would deliver the electricity to power and light the cities of the world. That same year, Edison founded the Edison Illuminating Company—the first investor-owned electric utility—which later became the General Electric Corporation. In 1881, he left Menlo Park to establish facilities in several cities where electrical systems were being installed. In 1882, the Pearl Street generating station provided 110 volts of electrical power to 59 customers in lower Manhattan.

Read More at:
https://www.biography.com/people/thomas-edison-9284349

T. A. EDISON.
Electric-Lamp.
898. Patented Jan. 27, 1880.

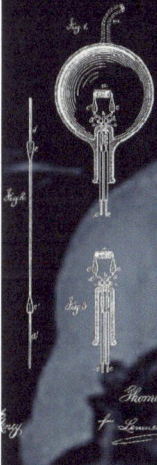

Inventor
Thomas A. Edison

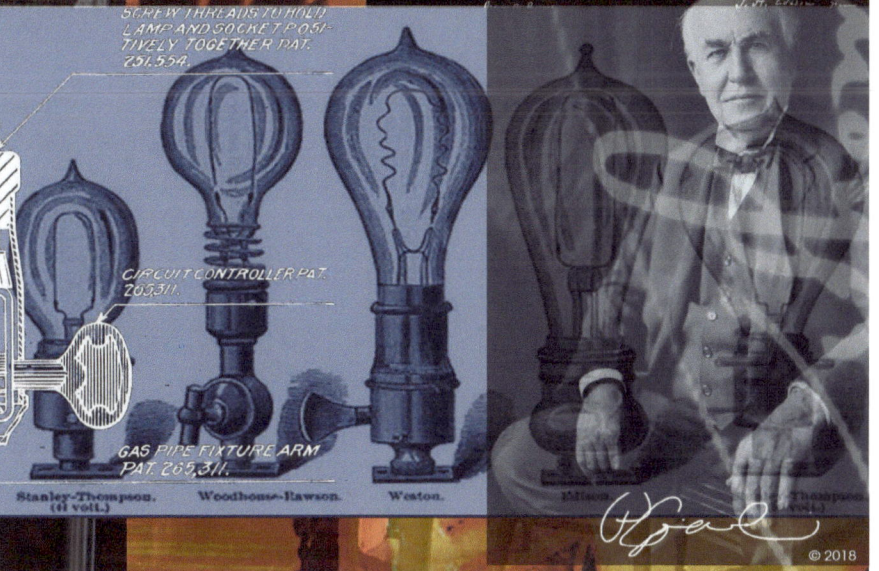

10.14.1878

edison

EXHAUSTED GLASS GLOBE
PAT. 223,898 -227,229
BLOWN FROM POT GLASS
PAT. 266,447.

HIGH RESISTANCE CARBON
FILAMENT PAT. 230,255
HEATED TO INCANDESCENCE
WHILE LAMP WAS BEING EX-
HAUSTED PAT. 265,777.

WIRES SEALED IN GLASS
PAT. 223,898 -227,229.

Thomas A. Edison

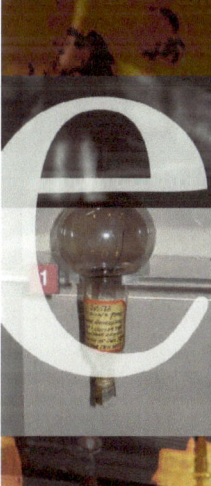

POINT WHERE TWO PARTS
ARE JOINED BY FUSION
PAT. 230,255.

BASE OF INSULATING MATER-
IAL WITH TWO CONTACTS
UPON IT PAT. 251,554 -265,311-
417,631 -264,737.

SOCKET CONTACTS CORRE-
SPONDING TO LAMP CON-
TACTS PAT. 251,554 -265,311.

DETACHABLE SOCKET PAT.
251,554 -265,311.

SCREW THREADS TO HOLD
LAMP AND SOCKET POSI-
TIVELY TOGETHER PAT.
251,554.

CIRCUIT CONTROLLER PAT.
265,311.

GAS PIPE FIXTURE ARM
PAT. 265,311.

Stanley-Thompson.
(4 volt.)

Woodhouse-Rawson.

Weston.

© 2018

Karl Ferdinand Braun
(1850 -1918)

Karl Ferdinand Braun - studied at the Universities of Marburg and Berlin and graduated in 1872 with a paper on the oscillations of elastic strings. He worked as assistant to Professor Quincke at Würzburg University and in 1874 accepted a teaching appointment to the St. Thomas Gymnasium in Leipzig. Two years later he was appointed Extraordinary Professor of Theoretical Physics at the University of Marburg, and in 1880 he was invited to fill a similar post at Strasbourg University. Braun was made Professor of Physics at the Technische Hochschule in Karlsruhe in 1883 and was finally invited by the University of Tübingen in 1885; one of his tasks there was to build a new Physics Institute. Ten years later, in 1895, he returned to Strasbourg as Principal of the Physics Institute, where he remained, in spite of an invitation from Leipzig University to succeed G. Wiedemann.

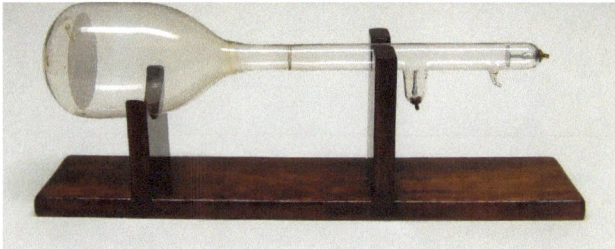

Braun's first investigations were concerned with oscillations of strings and elastic rods, especially with regard to the influence of the amplitude and environment of rods on their oscillations. Other studies were based on thermodynamic principles, such as those on the influence of pressure on the solubility of solids.

His most important works, however, were in the field of electricity. He published papers on deviations from Ohm's law and on the calculations of the electromotive force of reversible galvanic elements from thermal sources. His practical experiments led him to invent what is now called Braun's electrometer, and also a cathode-ray oscillograph, constructed in 1897.

Read More at:

https://www.nobelprize.org/nobel_prizes/physics/laureates/1909/braun-bio.html

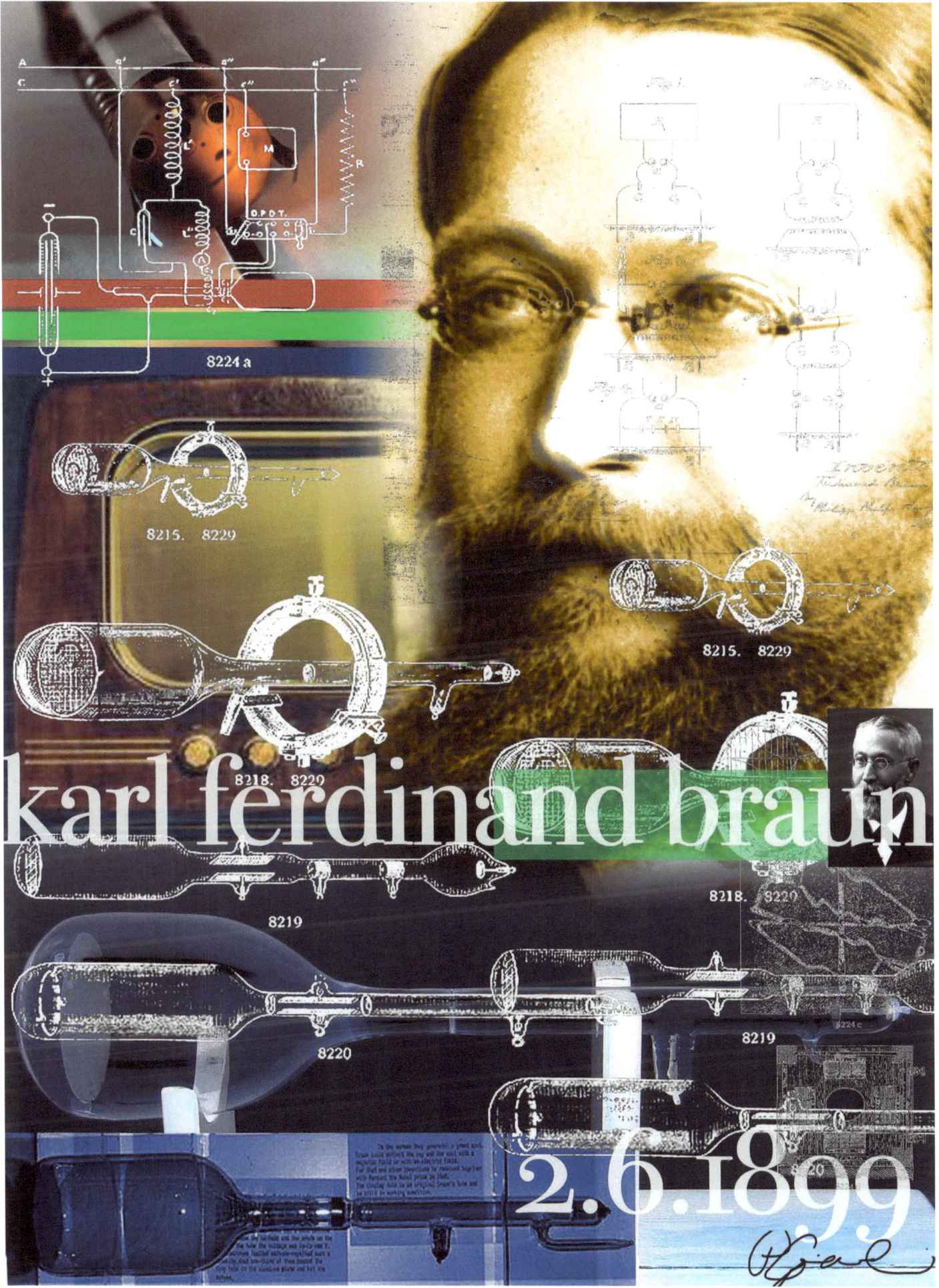

karl ferdinand braun

2.6.1899

Nikola Tesla

(1856 - 1943)

Nikola Tesla, (born July 9/10, 1856, Smiljan, Austrian Empire [now in Croatia]—died January 7, 1943, New York, New York, U.S.), Serbian American inventor and engineer who discovered and patented the rotating magnetic field, the basis of most alternating-current machinery. He also developed the three-phase system of electric power transmission. He immigrated to the United States in 1884 and sold the patent rights to his system of alternating-current dynamos, transformers, and motors to George

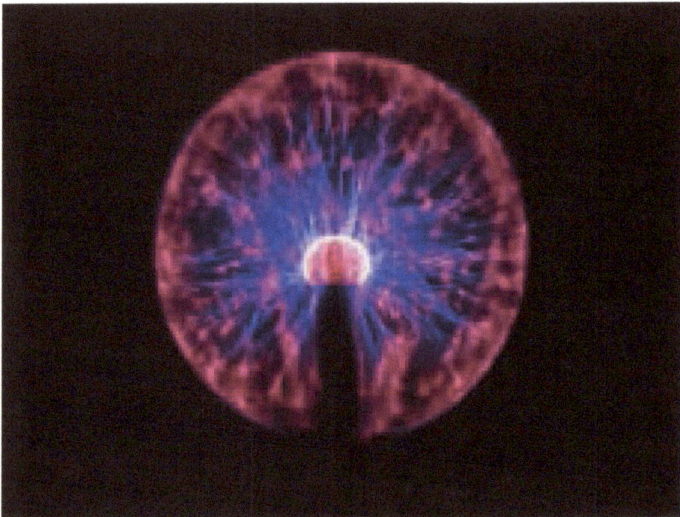

Credit: https://pxhere.com/en/photo/631744

Westinghouse. In 1891 he invented the Tesla coil, an induction coil widely used in radio technology.

Tesla was from a family of Serbian origin. His father was an Orthodox priest; his mother was unschooled but highly intelligent. As he matured, he displayed remarkable imagination and creativity as well as a poetic touch.

Training for an engineering career, he attended the Technical University at Graz, Austria, and the University of Prague. At Graz he first saw the Gramme dynamo, which operated as a generator and, when reversed, became

an electric motor, and he conceived a way to use alternating current to advantage. Later, at Budapest, he visualized the principle of the rotating magnetic field and developed plans for an induction motor that would become his first step toward the successful utilization of alternating current. In 1882 Tesla went to work in Paris for the Continental Edison Company, and, while on assignment to Strassburg in 1883, he constructed, after work hours, his first induction motor. Tesla sailed for America in 1884, arriving in New York with four cents in his pocket, a few of his own poems, and calculations for a flying machine. He first found employment with Thomas Edison, but the two inventors were far apart in background and methods, and their separation was inevitable.

In May 1888 George Westinghouse, head of the Westinghouse Electric Company in Pittsburgh, bought the patent rights to Tesla's polyphase system of alternating-current dynamos, transformers, and motors. The transaction precipitated a titanic power struggle between Edison's direct-current systems and the Tesla-Westinghouse alternating-current approach, which eventually won out.

Tesla soon established his own laboratory, where his inventive mind could be given free rein. He experimented with shadowgraphs similar to those that later were to be used by Wilhelm Röntgen when he discovered X-rays in 1895. Tesla's countless experiments included work on a carbon button lamp, on the power of electrical resonance, and on various types of lighting.

Read More at:

https://www.britannica.com/biography/Nikola-Tesla

January 7, 1943

tesla

Nikola Tesla

© 2018

IO

Philo T. Farnsworth
(1906 - 1971)

American inventor who developed the first all-electronic television system.

Farnsworth was a technical prodigy from an early age. An avid reader of science magazines as a teenager, he became interested in the problem of television and was convinced that mechanical systems that used, for example, a spinning disc would be too slow to scan and assemble images many times a second. Only an electronic system could scan and assemble an image fast enough, and by 1922 he had worked out the basic outlines of electronic television.

In 1923, while still in high school, Farnsworth also entered Brigham Young University in Provo, Utah, as a special student. However, his father's death in January 1924 meant that he had to leave Brigham Young and work to support his family while finishing high school.

Farnsworth had to postpone his dream of developing television. In 1926 he went to work for charity fund-raisers George Everson and Leslie Gorrell. He convinced them to go into a partnership to produce his television system.

Read More at:
https://www.britannica.com/biography/Philo-Farnsworth

Philo Farnsworth with his television

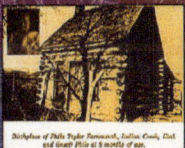

II

Alan Turing
(1912-1954)

Alan Turing, in full Alan Mathison Turing, (born June 23, 1912, London, England—died June 7, 1954, Wilmslow, Cheshire), British mathematician and logician, who made major contributions to mathematics, cryptanalysis, logic, philosophy, and mathematical biology and also to the new areas later named computer science, cognitive science, artificial intelligence, and artificial life.

Early Life And Career

The son of a civil servant, Turing was educated at a top private school. He entered the University of Cambridge to study mathematics in 1931. After graduating in 1934, he was elected to a fellowship at King's College (his college since 1931) in recognition of his research in probability theory. In 1936 Turing's seminal paper "On Computable Numbers, with an Application to the Entscheidungsproblem [Decision Problem]" was recommended for publication by the American mathematical logician Alonzo Church, who had himself just published a paper that reached the same conclusion as Turing's, although by a different method. Turing's method (but not so much Church's) had profound significance for the emerging science of computing. Later that year Turing moved to Princeton University to study for a Ph.D. in mathematical logic under Church's direction (completed in 1938).

Read More at:

https://www.britannica.com/biography/Alan-Turing

©

9.4.1939

TAPE

Offset print raised for a "mark"

Eraser

Offset-printing + eraser roller

Tractor roller

mark on tape

mark in process of erasure

HEAD

motors

erased electric eye looking at tape square

indexing hole

scanned symbol

Print, Erase
Left, Right

tractor hole

Print motor

Tractor motor

TABLE	Current state A			Current state B			Current state C		
	Write symbol	Move tape	Next state	Write symbol	Move tape	Next state	Write symbol	Move tape	Next state

turing

ALAN TURING
1912 - 1954
Founder of computer science and cryptographer, whose work was key to breaking the wartime Enigma codes, lived and died here.

Between 4 September 1939 and the summer of 1944, Alan Turing lodged at The Crown Inn, at Shenley Brook End, a village to the west of Bletchley.

ALA

Founder and crypto

was k

warti

live

12

Grace Murray Hopper
(1906-1992)

Born in New York City in 1906, Grace Hopper joined the U.S. Navy during World War II and was assigned to program the Mark I computer. She continued to work in computing after the war, leading the team that created the first computer language compiler, which led to the popular CO-BOL language. She resumed active naval service at the age of 60, becoming a rear admiral before retiring in 1986. Hopper died in Virginia in 1992.

Early Life

Born Grace Brewster Murray in New York City on December 9, 1906, Grace Hopper studied math and physics at Vassar College. After graduating from Vassar in 1928, she proceeded to Yale University, where, in 1930, she received a master's degree in mathematics. That same year, she married Vincent Foster Hopper, becoming Grace Hopper (a name that she kept even after the couple's 1945 divorce). Starting in 1931, Hopper began teaching at Vassar while also continuing to study at Yale, where she earned a Ph.D. in mathematics in 1934—becoming one of the first few woman to earn such a degree.

Read More at:

https://www.biography.com/people/grace-hopper-21406809

6.1.1944

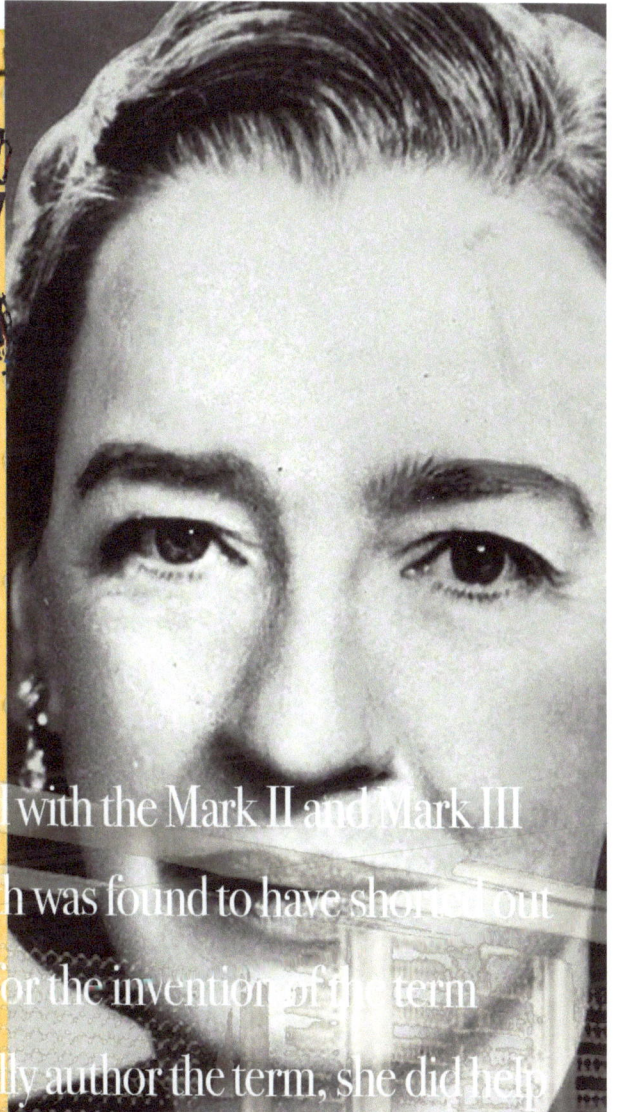

As a research fellow at Harvard, she worked with the Mark II and Mark III computers. She was at Harvard when a moth was found to have shorted out the Mark II, and is sometimes given credit for the invention of the term "computer bug"—though she didn't actually author the term, she did help popularize it.

murray
hopper

United States Navy Rear Admiral

13

Pierre Etienne Bezier
(1910-1999)

Pierre Etienne Bezier was born on September 1, 1910 in Paris. Son and grandson of engineers, he chose this profession too and enrolled to study mechanical engineering at the Ecole des Arts et Metiers and received his degree in 1930. In the same year he entered the Ecole Superieure d'Electricite and earnt a second degree in electrical engineering in 1931. In 1977, 46 years later, he received his DSc degree in mathematics from the University of Paris.

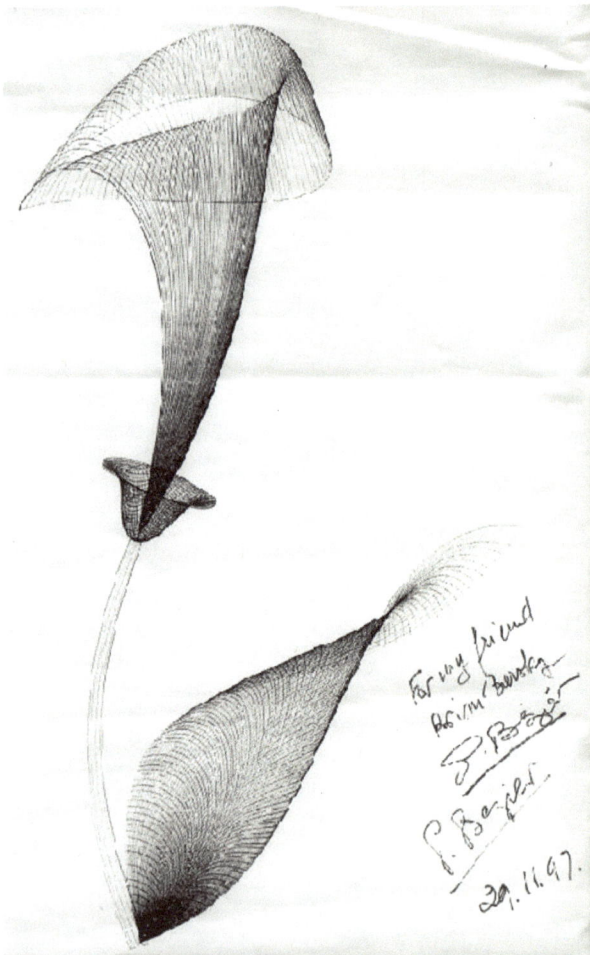

In 1933, aged 23, Bezier entered Renault and worked for this company for 42 years. He started as Tool Setter, became Tool Designer in 1934 and Head of the Tool Design Office in 1945. In 1948, as Director of Production Engineering he was responsable for the design of the transfer lines producing most of the 4 CV mechanical parts. In 1957, he became Director of Machine Tool Division and was responsable for the automatic assembly of mechanical components, and for the design and production of an NC drilling and milling machine, most probably one of the first machines in Europe. Bezier become managing staff member for technical development in 1960 and held this position until 1975 when he retired.

Read More at:

http://www.engology.com/eng5bezier.htm#selection-11.0-15.701

$$\binom{n+1}{i}(1-t)\mathbf{b}_{i,n} = \binom{n}{i}\mathbf{b}_{i,n+1} \Rightarrow (1-t)\mathbf{b}_{i,n}$$

$$\binom{n+1}{i+1}t\mathbf{b}_{i,n} = \binom{n}{i}\mathbf{b}_{i+1,n+1} \Rightarrow t\mathbf{b}_{i,n} =$$

$$\mathbf{B}(t) = (1-t)\sum_{i=0}^{n}\mathbf{b}_{i,n}(t)\mathbf{P}_i + t\sum$$

Bézier

9.1.00

Program Manager

$$\mathbf{B}(t) = \sum_{i=0}^{n}\binom{n}{i}(1-t)^{n-i}t^{i}\mathbf{P}_i$$

$$= (1-t)^{n}\mathbf{P}_0 + \binom{n}{1}(1-t)^{n-1}t\mathbf{P}_1$$

$$\cdots + \binom{n}{n-1}(1-t)t^{n-1}\mathbf{P}_{n-1} +$$

Gordon Moore

(1929-)

Gordon Moore, in full Gordon E. Moore, (born January 3, 1929, San Francisco, California), American engineer and cofounder, with Robert Noyce, of Intel Corporation.

From Shockley To Intel

Moore was particularly excited about the potential of the transistor, a recent invention awaiting the development of practical manufacturing techniques. In 1956 Moore returned to California to work at Shockley Semiconductor Laboratory, which William Shockley, one of the Nobel Prize-winning inventors of the transistor, had just opened in Palo Alto. The new laboratory was researching manufacturing methods for silicon-based transistors, but after a hectic year-and-a-half under Shockley's management—capped by an appeal by Moore and others that the company hire a professional manager—Moore and seven colleagues resigned and joined with Fairchild Camera and Instrument Corporation to form a new company, Fairchild Semiconductor Corporation, in Santa Clara.

Read More at:

https://www.britannica.com/biography/Gordon-Moore

Credit: Carnegie Mellon University, Department of Civil and Environmental Engineering

Moore's Law

In 1965, Intel co-founder Gordon Moore predic...
of transistors on a piece of silicon wo...
years—in in... la... dubbed "Mo...
old true, ...e-shrinking transisto...
10 Bil...
...chip in the number of transistors on a ...

$$R1\ 27K \quad R2\ 15K \quad R3\ 1K \quad R4\ 15K \quad +9V$$
Q1 2N2646 \quad Q2 2N2646

B2 \quad B2

E \quad E

B1 \quad B1

C1 20µF \quad C2 .05µF

SPKR

D1 1N98 \quad 1K

FIGURE 8

The Electrochemical Society
Gordon E. Moore Medal

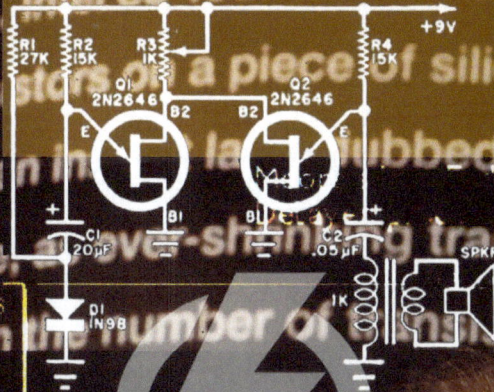

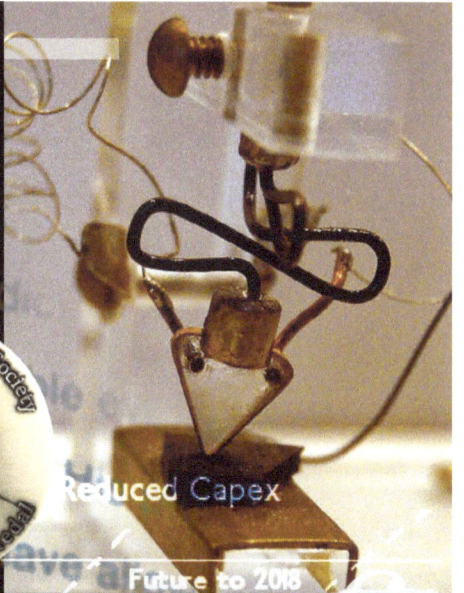

Reduced Capex

Future to 2018

4.19.1965

EVERY 18 MONTHS
$2300 \times 2^{0.6667\,(Year - 1971)}$

00 MILLION
2SC1970
13521

10 MILLION

1 MILLION

TRANSISTORS / CHIP DOUBLING EVERY 2 YEARS
(WORKED OUT WELL BUT NOT MUCH LONGER)
$N = 2300 \times 2^{0.5\,(Year - 1971)}$

PENTIUM

486

386

286

100,000

Based on logistic regression, asymptote at 6.25 billion.

...TANIUM

NATIONAL MEDAL OF TECHNOLOGY AND INNOVATION

(intel)®

moore

086

,000

8080

2300 \quad 80

40

By Clayton Hallmark
Dedicated
Professor Fr... ... Terman

1971 \quad 980 \quad Year \quad 1990 \quad 2000 \quad 2004 \quad 201... \quad 2015 \quad 2018

898 1900 1902 1905 1906 1908 1910 1912 1919 1921 1930 1932 1939 1940 1945 1946 1947 1948 1949 1950
e⁻ quantum effect
theory of solids
P N

951 1952 1953 1954 1955 1956 1957 1958 1960 1961 1965 1967 1968 1971 1972 1975 1981 1989 1997
Bell Labs \quad Hearing Aids \quad traitorous 8 \quad Mini-computer \quad Moore's Law \quad Fairchild Semiconductor \quad Microsoft \quad www

© 2018

15 Douglas Engelbart
(1925 - 2013)

Born on January 30, 1925, in Portland, Oregon, Douglas C. Engelbart was a pioneer in the design of interactive computer environments who invented the computer mouse in 1964. He also created the first two-dimensional editing system, and was the first to demonstrate the use of mixed text-graphics and shared-screen viewing. He was director of SRI International's Augmentation Research Center in Palo Alto and founded Stanford University's Bootstrap Project. Engelbart died in Atherton, California, on July 2, 2013, at age 88.

Douglas C. Engelbart went on to earn a Ph.D. in electrical engineering from the University of California, Berkeley, in 1955. After returning to the school for a stint as an acting assistant professor, Engelbart began a career at the Stanford Research Institute (later renamed SRI International). Around this same time, he began focusing on an approach that he termed "bootstrapping," in which he asserted the fields of engineering and science would be greatly improved if computer power were shared among researchers.

In the early 1960s, Engelbart founded SRI International's Augmentation Research Center in Palo Alto in an effort to further research information processing and computer-sharing tools and methods. Soon after, Engelbart designed and was the primary developer of the oN-Line System, also known as NLS, a revolutionary computer-sharing system.

In 1964, Engelbart conceptualized and created the first design for the computer mouse. While Engelbart believed that the point-and-click computer device could be equipped with up to 10 buttons, the first mouse would have just three. The inventor went on to create the first two-dimensional editing system, and was the first to demonstrate the use of mixed text-graphics and shared-screen viewing.

Engelbart served as director of the Augmentation Research Center from its inception until 1977. The center was transferred to Tymshare in 1978, with NLS being renamed "Augment. In 1989, Engelbart founded the Bootstrap Project at Stanford University.

Read More at:
https://www.biography.com/people/douglas-c-engelbart-9287574

Credit: APIC/Getty Images

16

Vint Cerf
(1943 -)

At Google, Vint Cerf contributes to global policy development and continued spread of the Internet. Widely known as one of the "Fathers of the Internet," Cerf is the co-designer of the TCP/IP protocols and the architecture of the Internet. He has served in executive positions at the Internet Society, the Internet Corporation for Assigned Names and Numbers, the American Registry for Internet Numbers, MCI, the Corporation for National Research Initiatives and the Defense Advanced Research Projects Agency and on the faculty of Stanford University.

Vint Cerf sits on US National Science Board and is a Visiting Scientist at the Jet Propulsion Laboratory. Cerf is a Fellow of the IEEE, ACM, American Association for the Advancement of Science, American Academy of Arts and Sciences, British Computer Society, Worshipful Company of Information Technologists, Worshipful Company of Stationers and is a member of the National Academy of Engineering.

© Klaus Tschira Stiftung / Peter Badge

Cerf is a recipient of numerous awards and commendations in connection with his work on the Internet, including the US Presidential Medal of Freedom, US National Medal of Technology, the Queen Elizabeth Prize for Engineering, the Prince of Asturias Award, the Japan Prize, the Charles Stark Draper award, the ACM Turing Award, the Legion d'Honneur and 25 honorary degrees.

Read More at:

https://royalsociety.org/people/vint-cerf-12851/

Ted Nelson

(1940 -)

Theodor Holm "Ted" Nelson (born June 17, 1937) is an American pioneer of information technology, philosopher, and sociologist. He coined the terms hypertext and hypermedia in 1963 and published them in 1965.[1]

Nelson earned a B.A. in philosophy from Swarthmore College in 1959. While there, he made an experimental humorous student film titled The Epiphany of Slocum Furlow, in which the titular hero discovers the meaning of life. His contemporary at the college, musician and composer Peter Schickele, scored the film.[6] Following a year of graduate study in sociology at the University of Chicago, Nelson began graduate work in philosophy at Harvard University in 1960, ultimately earning an A.M. in sociology from the Department of Social Relations in 1963. During his graduate studies, Nelson was a photographer and filmmaker at John C. Lilly's Communication Research Institute in Miami, Florida, where he briefly shared an office with Gregory Bateson. From 1964 to 1966, he was an instructor in sociology at Vassar College.[7]

During college and graduate school, he envisioned a computer-based writing system that would provide a lasting repository for the world's knowledge, and also permit greater flexibility of drawing connections between ideas. This came to be known as Project Xanadu.

Much later in life, he obtained a Ph.D. in media and governance from Keio University in 2002.

Read More at:

http://www.computerhistory.org/fellowawards/hall/john-warnock/

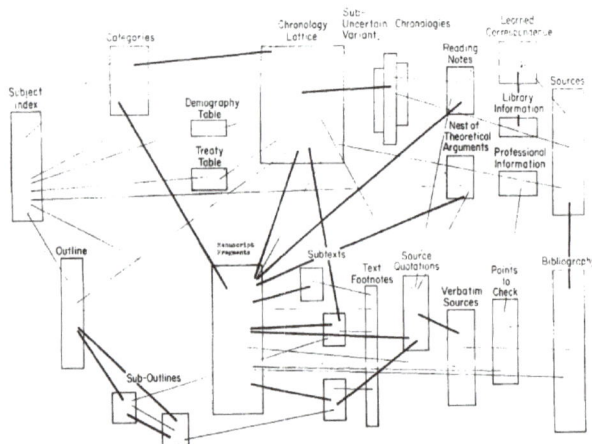

FIGURE 4—ELF's capacity for total filing: hypothetical use by historian. Thin lines indicate links; heavy rules indicate some of same entries.

© Ed Kashi/CORBIS

John Warnock

(1940 -)

John Warnock was born in Salt Lake City, Utah, in 1940. He holds a B.S. in mathematics and philosophy (1961), an M.S. in mathematics (1964), and a Ph.D. in electrical engineering (1969), all from the University of Utah.

Warnock joined Xerox's Palo Alto Research Center (PARC) in 1978, where he was a principal scientist in their computer sciences laboratory working on interactive graphics. As part of his responsibilities, Warnock, along with colleague Chuck Geschke and others, developed a printer protocol called Interpress.

When Xerox declined to pursue the Interpress idea further, Warnock and Geschke started their own company, Adobe. Their work evolved into the PostScript page description language, which, when combined in 1985 with hardware from Apple Computer, formed the first "desktop publishing" (DTP) system. DTP systems allowed nearly anyone to electronically compose documents and print them as they appeared on the screen. This new approach to electronic printing allowed business users to greatly improve the quality and efficiency of their document production.

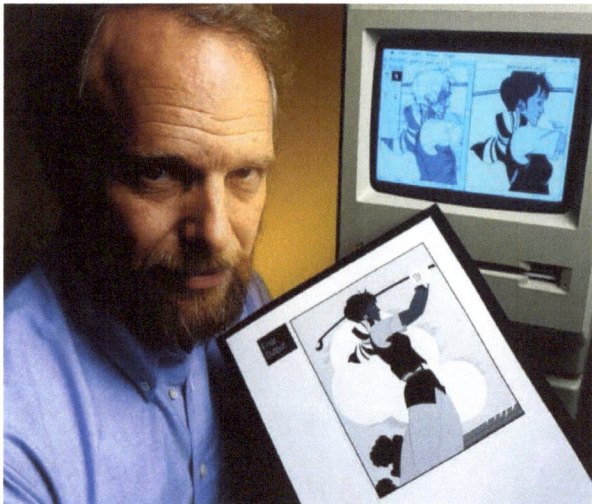

Read More at:
http://www.computerhistory.org/fellowawards/hall/john-warnock/

© Ed Kashi/CORBIS

19

Adele Goldberg

(1945 -)

A computer scientist who participated in the development of the programming language Smalltalk-80 and various concepts related to object oriented programming while a researcher at the Xerox Palo Alto Research Center, PARC, in the 1970s. She eventually became manager of the System Concepts Laboratory where she, Alan Kay, and others developed Smalltalk-80. She and Kay also were involved in the development of design templates, forerunners of the design patterns commonly used in software design. In 1988 she left PARC to co-found ParcPlace Systems, a company that created development tools for Smalltalk-based applications. She served as president of the Association for Computing Machinery (ACM) from 1984 to 1986, and, together with Alan Kay and Dan Ingalls, received the ACM Software Systems Award in 1987. She also received PC Magazine's Lifetime Achievement Award in 1990. In 1994 she was inducted as a Fellow of the Association for Computing Machinery. Many of the concepts she and her team developed at PARC became the basis for graphically based user interfaces, replacing the earlier command line based systems. She has reported that, Steve Jobs demanded a demonstration of the Smalltalk System, which she refused to give him. Her superiors eventually ordered her to comply, and she did, satisfied that the decision to "give away the kitchen sink" to Jobs and his team was then their responsibility. Apple eventually used many of the ideas in the Alto and their implementations as the basis for their Macintosh desktop. She is currently working for Neometron, Inc., of Palo Alto, California.

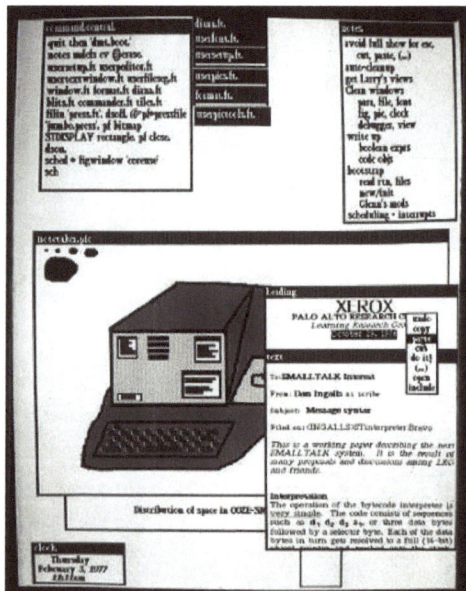

Smalltalk screenshot – © PARC (Palo Alto Research Center, Incorporated)

Read More at:

https://www.ithistory.org/honor-roll/dr-adele-goldberg

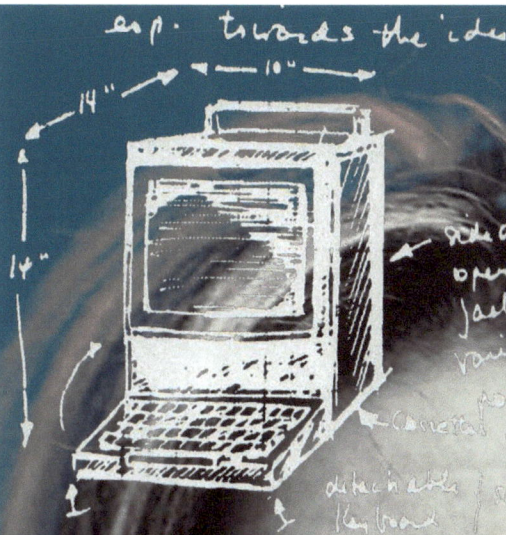

Dr. Goldberg has served as president of the ACM and has received numerous awards for her contributions to computing. Along with Alan Kay, Dr. Goldberg developed Smalltalk and wrote much of the documentation. Smalltalk was used to prototype the WIMP (windows, icons, menus, pointers) interface at Xerox Parc, the cornerstone for today's modern graphical user interfaces.

goldberg

9.1.1973

```
[filename, pathname] = uigetfile('*.avi','Pick AVI file');
if ~isequal(filename, 0)
try
    handles.mov = aviread(fullfile(pathname,filename));
    guidata(hObject, handles); % Update handles structure
    % display 1st frame
    subplot(handles.input_frame);
    image(rgb2gray(handles.mov(1).cdata));
```

20

Steve Wozniak
(1950 -)

Steve Wozniak was born in San Jose, California, on August 11, 1950. In partnership with his friend Steve Jobs, Wozniak invented the Apple I computer. The pair founded Apple Computers in 1976 with Ronald Wayne, releasing some of the first personal computers on the market. Wozniak also personally developed the next model, Apple II, which established Apple as a major player in microcomputing.

© Apple-1 at Smithsonian

Founding Apple Computer

The son of an engineer at Lockheed Martin, Stephen Gary Wozniak, born on August 11, 1950, was fascinated by electronics at an early age. Although he was never a star student in the traditional sense, Wozniak had an aptitude for building working electronics from scratch.

During his brief stint at the University of California at Berkeley, Steve Wozniak met Steve Jobs, who was still in high school, through a mutual friend. The two later paired up to form Apple Computer on April 1, 1976, prompting Wozniak to quit his job at Hewlett-Packard.

Working out of a family garage, he and Jobs attempted to produce a user-friendly alternative to the computers that were being introduced by International Business Machines at that time. Wozniak worked on the invention of products, and Jobs was responsible for marketing.

Read More at:
https://www.biography.com/people/steve-wozniak-9537334

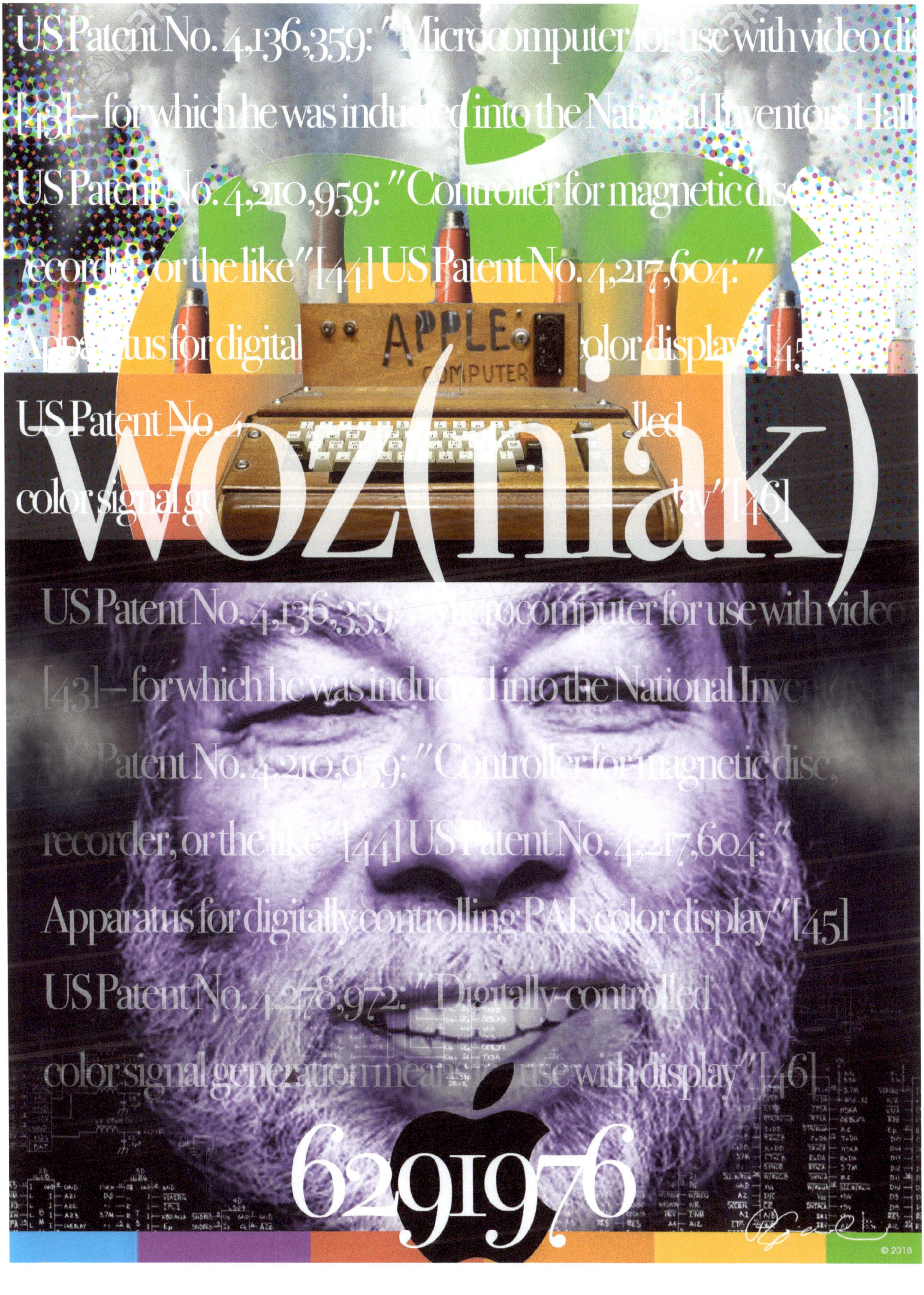

woz(niak)

US Patent No. 4,136,359: "Microcomputer for use with video dis[play]

[43] – for which he was inducted into the National Inventors Hall

US Patent No. 4,210,959: "Controller for magnetic disc

recorder, or the like" [44] US Patent No. 4,217,604: "

Apparatus for digital[ly] color display" [45

US Patent No. [controlled]

color signal g[enerator] [ay" [46]

US Patent No. 4,136,359: [Mi]crocomputer for use with video

[43]–for which he was inducted into the National Inven[tors]

[US] Patent No. 4,210,959: "Controller for magnetic disc,

recorder, or the like" [44] US Patent No. 4,217,604: "

Apparatus for digitally controlling PAL color display" [45]

US Patent No. 4,278,972: "Digitally-controlled

color signal generation means for use with display" [46]

6291976

© 2018

21

Jim Clark

(1944 -)

James Clark was born on March 23, 1944, in Plainview, Texas. At Stanford, he developed a program called the Geometry Engine. He started Silicon Graphics Inc. in 1981 before quitting the company in 1994. The same day, he and Marc Andreessen agreed to launch Netscape. It was sold to America Online in 1999, but Clark had already started a new venture by then, Healtheon, which would soon merge with WebMD.

Early Life

From his poverty-stricken childhood in Texas, Jim Clark rose to become one of the most famous and successful high-tech entrepreneurs in the world. Born in 1944, Clark was a strong-willed boy with a rebellious streak that became more pronounced after his parents divorced when he was 14. His mother supported him and his two siblings on $225 a month (she now lives in a house bought with Netscape stock that her son gave her).

Entrepreneurial Success

At Stanford, he developed a program called the Geometry Engine which generated three-dimensional computer graphics. He left academia in 1981 to start Silicon Graphics Incorporated with $25,000 he borrowed from a friend. By 1986, the company had revenues of $40 million and had revolutionized the design process for everything from bridges and airplanes to special effects for movies, including Terminator 2 (1991) and Jurassic Park (1993).

As the company grew, shareholders brought in more conservative management who put the reins on Clark's freewheeling style. Clark, rumored to have an explosive temper, felt stifled. In 1994, he quit the company and sold his stock.

© SGI

Read More at:

https://www.biography.com/people/james-clark-9542204

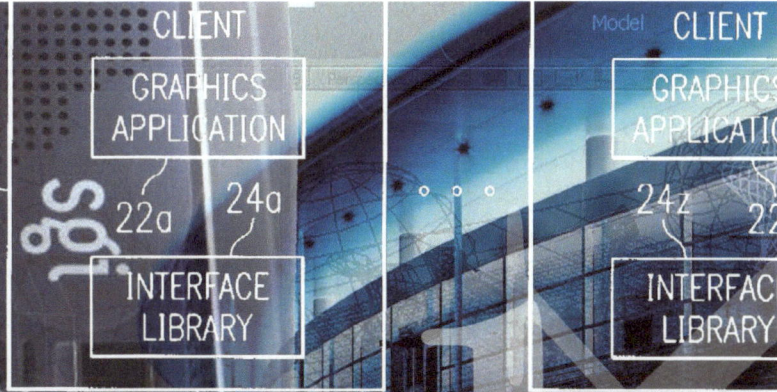

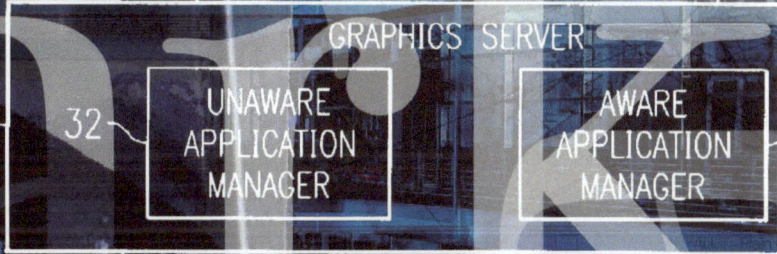

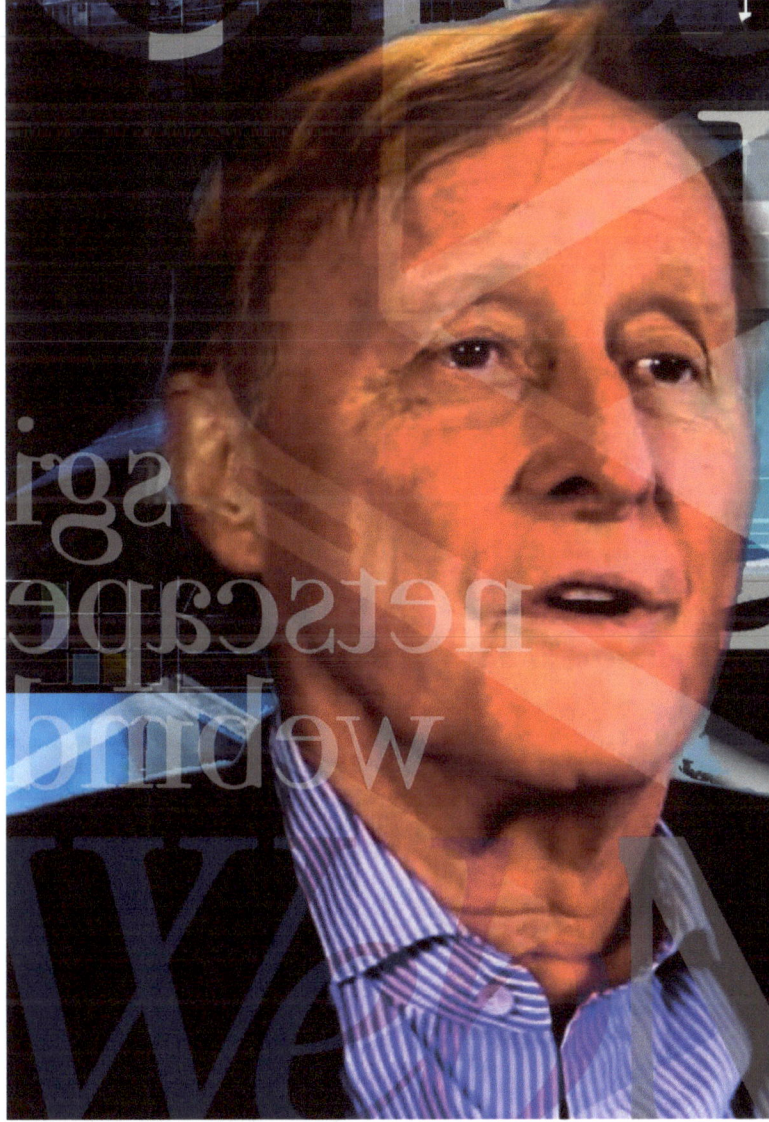

John Knoll

(1962 -)

John Knoll (born October 6, 1962) is an American visual effects supervisor and chief creative officer (CCO) at Industrial Light & Magic (ILM). [1] One of the original creators of Adobe Photoshop (along with his brother, Thomas Knoll), he has also worked as visual effects supervisor on the Star Wars prequels and the 1997 special editions of the original trilogy. He also served as ILM's visual effects supervisor for Star Trek Generations and Star Trek: First Contact, as well as the Pirates of the Caribbean series. Along with Hal Hickel, Charles Gibson and Allen Hall, Knoll and the trio's work on Pirates of the Caribbean: Dead Man's Chest earned them the Academy Award for Best Visual Effects.[2]

Knoll has been praised by directors James Cameron, Gore Verbinski, Guillermo del Toro, and Brad Bird. Del Toro, who worked with Knoll for the first time on Pacific Rim, stated "He basically has the heart of a kid and the mind of a scientist, and that's a great combination."[3]

Knoll is also the inventor of Knoll Light Factory, a lens flare generating software inspired by his work at Industrial Light and Magic.

Read More at:

https://en.wikipedia.org/wiki/John_Knoll

23

Sir Tim Berners-Lee
(1955 -)

The inventor of the World Wide Web and one of Time Magazine's '100 Most Important People of the 20th Century', Sir Tim Berners-Lee is a scientist and academic whose visionary and innovative work has transformed almost every aspect of our lives.

Having invented the Web in 1989 while working at CERN and subsequently working to ensure it was made freely available to all, Berners-Lee is now dedicated to enhancing and protecting the web's future. He is a Founding Director of the World Wide Web Foundation, which seeks to ensure the web serves humanity by establishing it as a global public good and a basic right. He is also Director of the World Wide Web Consortium, a global web standards organisation he founded in 1994 to lead the web to its full potential. In 2012 he co-founded the Open Data Institute (ODI) which advocates for Open Data in the UK and globally. Sir Tim has advised a number of governments and corporations on ongoing digital strategies. A graduate of Oxford University, Sir Tim presently holds academic posts at the Massachusetts Institute of Technology at CSAIL (Computer Science and Artificial Intelligence Lab), (USA) and the University of Oxford (UK).

Sir Tim has received multiple accolades in recent years. These include receiving the first Queen' Elizabeth Prize for Engineering in 2013, election as a Fellow of the American Academy of Arts and Sciences in 2009 and being knighted by H.M. Queen Elizabeth in 2004. He has received over 10 honorary doctorates, is a member of the Internet Hall of Fame, and was awarded the Finland Millennium Prize in 2004, and the A.M. Turing Award — often called 'computing's Nobel Prize' — in 2016. In 2007, Berners-Lee was awarded the UK's Order of Merit – a personal gift of the monarch limited to just 24 living recipients. In 2012, he played a starring role in the opening ceremony for the Olympics, where, in front of an audience of some 900 million, he tweeted: "This is for everyone".

Read More at:
https://webfoundation.org/about/sir-tim-berners-lee/

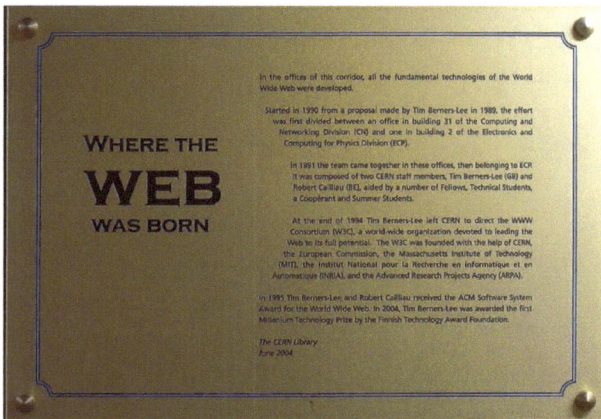

© CERN

WHERE THE WEB WAS BORN

24

Marc Andreessen

(1971 -)

Marc Andreessen, an American inventor, was born in Iowa and grew up in Wisconsin. While studying at the University of Illinois-Champagne, Andreessen became interested in the newly created Internet and started a graphic based browser to search the web. The company he created, Netscape, went public in 1995, but immediately began the "browser wars" with Microsoft.

Early Life

Inventor, Businessman. Born July 9, 1971 in Cedar Falls, Iowa, Marc Andreessen, co-founder of Netscape and inventor of the graphical Web browser, got bored with computers before he even made it through high school. Although the tall, blond boy, who grew up in New Lisbon, Wisconsin, taught himself BASIC programming from a library book at age nine, by high school he'd run out of things to do with the his TRS-80, one of the primitive personal computers available in the mid-1980s. At the University of Illinois in Champaign, Andreessen only majored in computer science because he wasn't doing well in electrical engineering. Even with his new major, he frequently skipped class or dozed off, he later claimed.

Creation of Mosaic

While working in a physics lab at college, Andreessen felt his old interest in computers rekindled when he noticed scientists sharing their work with other universities via the Internet in the early 1990s.

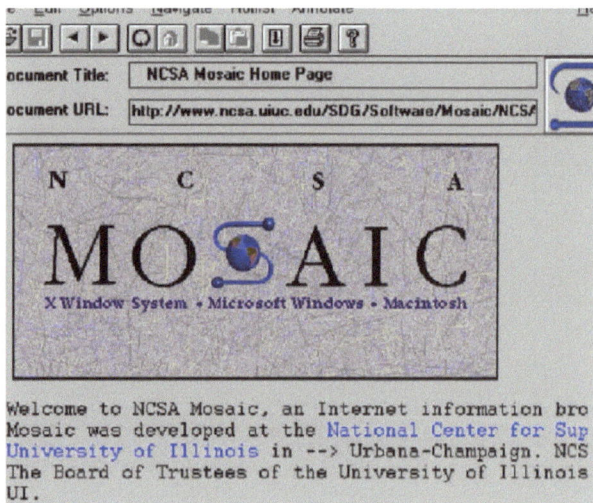

© Ed Kashi/CORBIS

Read More at:

https://www.biography.com/people/marc-andreessen-9542208

```
anchor {
e "http://home.netscape.com"
cription "Netscape"
DEF NetscapeN Separator {
    Coordinate3 {
        point [ -26.981634 -30.467456 97.903259,
                -20.832026 -30.467356 98.528000,
                -19.008020 -30.467 56 100.063354,
                -19.008020 -30.467 56 143.856140,
                -20.83202 -30.467 56 144.992767,
                -20.27320 -30.46  45.466705,
```

MOSAIC

X Window System • Windows • Macintosh

andreessen

netscape.wrl - WordPad

Edit View Insert Format Hel

ier New

ewsupdates

• • • ! • • • 1

Anchor F1{
e "http://home.netscape.com"
cription "Netscape"
DEF Netscape Separat
Coordinate3 {
point [-26.98   -30.467   97.90
        -20.832   -30.467   .5230
        -19.0008   -30.4674   56
        -19.0008020 -30.467
        -20.832026 -30.467
        -27.27320   -31.46

1.3.93

4

25

Jeff Bezos

(1964 -)

Jeff Bezos was born on January 12, 1964, in Albuquerque, New Mexico, to a teenage mother, Jacklyn Gise Jorgensen, and his biological father, Ted Jorgensen. The Jorgensens were married less than a year, and when Bezos was 4 years old his mother re-married, to Cuban immigrant Mike Bezos.

As a child, Jeff Bezos showed an early interest in how things work, turning his parents' garage into a laboratory and rigging electrical contraptions around his house. He moved to Miami with his family as a teenager, where he developed a love for computers and graduated valedictorian of his high school. It was during high school that he started his first business, the Dream Institute, an educational summer camp for fourth, fifth and sixth graders.

Bezos pursued his interest in computers at Princeton University, where he graduated summa cum laude in 1986 with a degree in computer science and electrical engineering. After graduation, he found work at several firms on Wall Street, including Fitel, Bankers Trust and the investment firm D.E. Shaw. It was there he met his wife, Mackenzie, and became the company's youngest vice president in 1990.

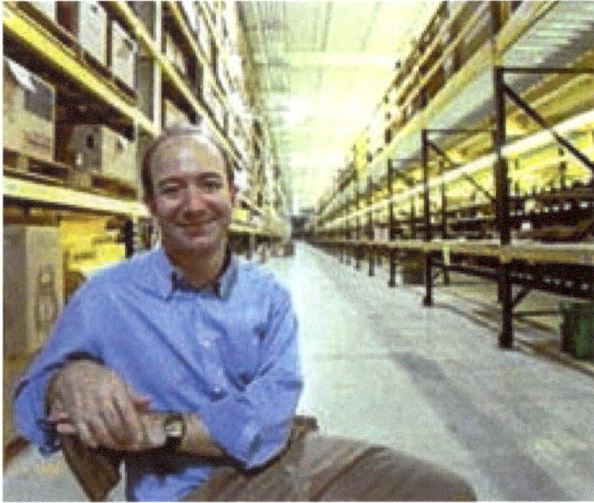

While his career in finance was extremely lucrative, Bezos chose to make a risky move into the nascent world of e-commerce. He quit his job in 1994, moved to Seattle and targeted the untapped potential of the internet market by opening an online bookstore.

Launching Amazon.com
Bezos set up the office for his fledgling company in his garage where, along with a few employees, he began developing software. They expanded operations into a two-bedroom house, equipped with three Sun Microstations, and eventually developed a test site. After inviting 300 friends to beta test the site, Bezos opened Amazon.com, named after the meandering South American River, on July 16, 1995.

Read More at:
https://www.biography.com/people/jeff-bezos-9542209

© David Burnett for TIME

bezos

7.16.1995

© 2018

Larry Page & Sergey Brin

(1973 -) & (1973 -)

Larry Page and Sergey Brin founded Google, the Internet search engine, while they were graduate students at Stanford University in Palo Alto, California. Since its founding in 1998, Google has become one of the most successful dot-com businesses in history. Both Page and Brin were reluctant entrepreneurs who were committed to developing their company on their own terms, not those dictated by the prevailing business culture.

Not instant best friends

Page grew up in the East Lansing, Michigan, area, where his father, Carl Victor Page, was a professor of computer science at Michigan State University. The senior Page was also an early pioneer in the field of artificial intelligence, and reportedly gave his young son his first computer when Larry was just six years old. Several years later Page entered the University of Michigan, where he earned an undergraduate degree in engineering with a concentration in computer engineering.

His first jobs were at Advanced Management Systems in Washington, D.C., and then at a company called CogniTek in Evanston, Illinois. An innovative thinker with a sense of humor as well, Page once built a working ink-jet printer out of Lego blocks. He was eager to advance in his career, and decided to study for a Ph.D degree. He was admitted to the prestigious doctoral program in computer science at Stanford University. On an introductory weekend at the Palo Alto campus that had been arranged for new students, he met Sergey Brin. A native of Moscow, Russia, Brin was also the son of a professor, and came to the United States with his family when he was six. His father taught math at the University of Maryland, and it was from that school's College Park campus that Brin earned an undergraduate degree in computer science and math.

Brin was already enrolled in Stanford's PhD program when Page arrived in 1995. As Brin explained to Robert McGarvey of Technology Review, "I was working on data mining, the idea of taking large amounts of data, analyzing it for patterns and trying to extract relationships that are useful." One weekend Brin was assigned to a team that showed the new doctoral students around campus, and Page was in his group. Industry lore claims they argued the whole time, but soon found themselves working together on a research project. That 1996 paper, "Anatomy of a Large-Scale Hypertextual Web Search Engine," became the basis for the Google search engine.

Read more: http://www.notablebiographies.com/news/Ow-Sh/Page-Larry-and-Brin-Sergey.html#ixzz5FqodfQkZ

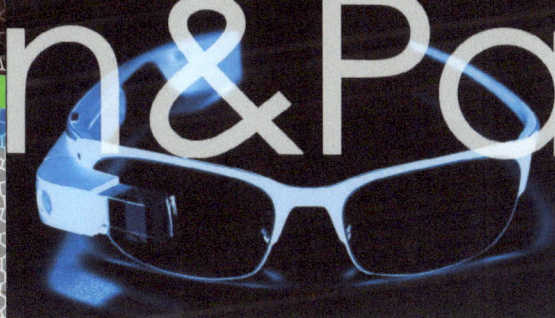

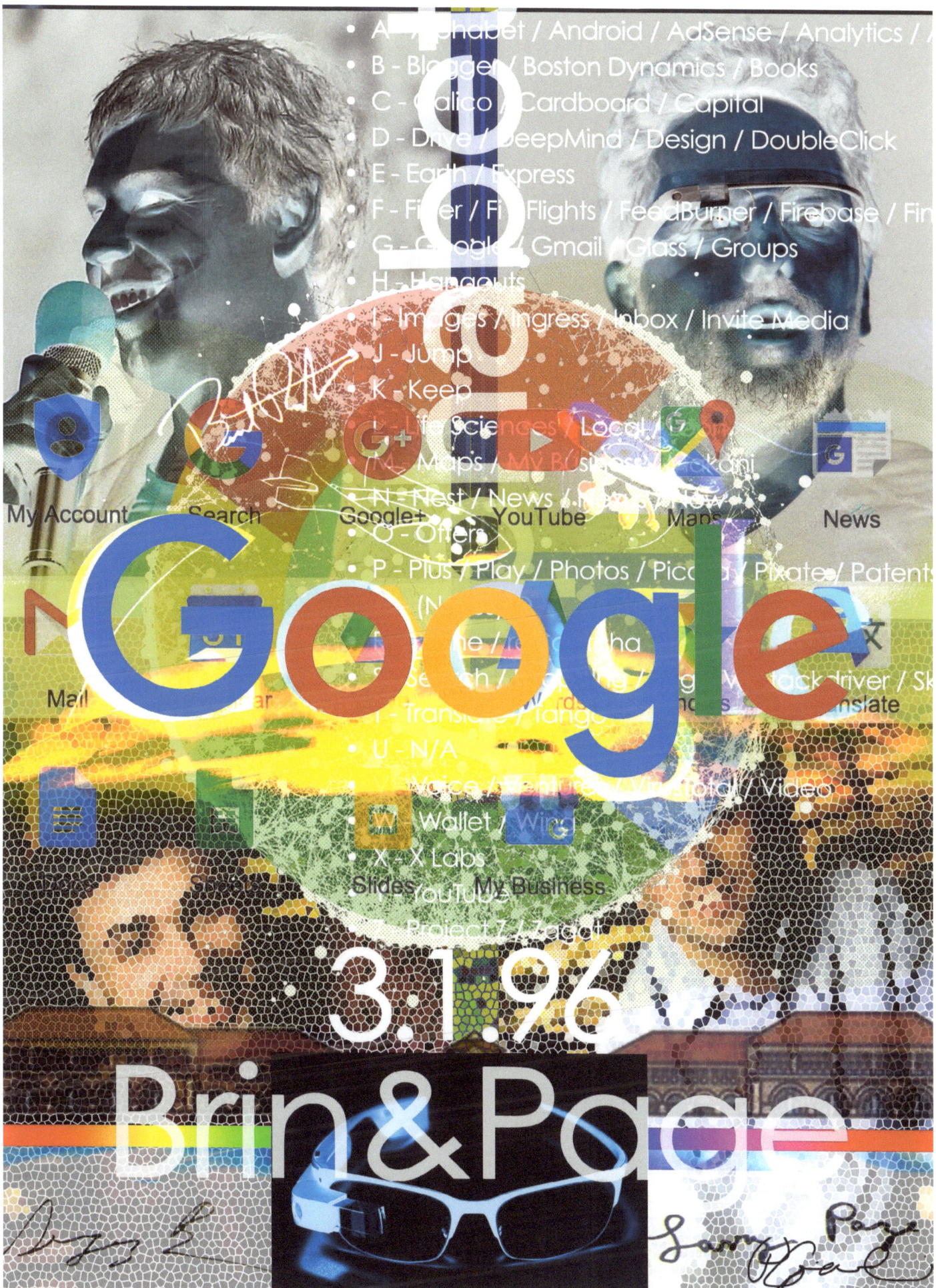

27

The "PayPal" Mafia
(1940 -)

"PayPal Mafia" is a term used to indicate a group of former PayPal employees and founders who have since founded and developed additional technology companies[1] such as Tesla Motors, LinkedIn, Palantir Technologies, SpaceX, YouTube, Yelp, and Yammer.[2] Most of the members attended Stanford University or University of Illinois at Urbana–Champaign at some point in their studies. Six members, Peter Thiel, Elon Musk, Reid Hoffman, Luke Nosek, Ken Howery, and Keith Rabois, have become billionaires.

History

Originally, PayPal was a money-transfer service offered by a company called Confinity which was acquired by X.com in 1999. Later X.com was renamed PayPal and purchased by eBay in 2002.[3][4] The original PayPal employees had difficulty adjusting to eBay's more traditional corporate culture and within four years all but 12 of the first 50 employees had left.[5] They remained connected as social and business acquaintances,[5] and a number of them worked together to form new companies in subsequent years. This group of PayPal alumni became so prolific that the term PayPal Mafia was coined.[3] The term[4][6] gained even wider exposure when a 2007 article in Fortune magazine used the phrase in its headline and featured a photo of former PayPal employees in gangster attire.[4][7][8][9]

Read More at:

https://en.wikipedia.org/wiki/PayPal_Mafia

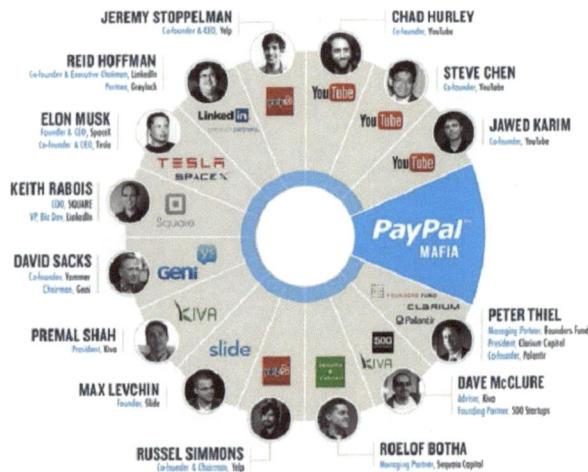

© Fortune Magazine

1999-2002 PayPal

28

Mark Zuckerberg

(1984 -)

Born on May 14, 1984, in White Plains, New York, Mark Zuckerberg co-founded the social-networking website Facebook out of his college dorm room. He left Harvard after his sophomore year to concentrate on the site, the user base of which has grown to more than 2 billion people, making Zuckerberg a billionaire. The birth of Facebook was portrayed in the 2010 film The Social Network.

Early Life

Mark Elliot Zuckerberg was born on May 14, 1984, in White Plains, New York, into a comfortable, well-educated family, and raised in the nearby village of Dobbs Ferry. His father, Edward Zuckerberg, ran a dental practice attached to the family's home. His mother, Karen, worked as a psychiatrist before the birth of the couple's four children—Mark, Randi, Donna and Arielle.

Zuckerberg developed an interest in computers at an early age; when he was about 12, he used Atari BASIC to create a messaging program he named "Zucknet." His father used the program in his dental office, so that the receptionist could inform him of a new patient without yelling across the room. The family also used Zucknet to communicate within the house. Together with his friends, he also created computer games just for fun. "I had a bunch of friends who were artists," he said. "They'd come over, draw stuff, and I'd build a game out of it."

Read More at:
https://www.biography.com/people/mark-zuckerberg-507402

Bibliography

https://www.biography.com/people/isaac-newton-9422656

https://www.britannica.com/biography/Michael-Faraday

http://www.computerhistory.org/babbage/engines/

http://www.computerhistory.org/babbage/adalovelace/

http://www.sciencemuseum.org.uk/onlinestuff/People/James_Clerk_Maxwell_183179.aspx

https://www.nobelprize.org/nobel_prizes/physics/laureates/1921/einstein-bio.html

https://www.nobelprize.org/nobel_prizes/physics/laureates/1909/braun-bio.html

https://www.britannica.com/biography/Nikola-Tesla

https://www.britannica.com/biography/Gordon-Moore

https://www.britannica.com/biography/Philo-Farnsworth

https://www.britannica.com/biography/Alan-Turing

https://www.biography.com/people/grace-hopper-21406809

http://www.engology.com/eng5bezier.htm#selection-11.0-15.701

https://www.britannica.com/biography/Gordon-Moore

https://www.biography.com/people/douglas-c-engelbart-9287574

https://royalsociety.org/people/vint-cerf-12851/

http://www.computerhistory.org/fellowawards/hall/john-warnock/

http://www.computerhistory.org/fellowawards/hall/john-warnock/

https://www.ithistory.org/honor-roll/dr-adele-goldberg

https://www.biography.com/people/steve-wozniak-9537334

https://www.biography.com/people/james-clark-9542204

https://en.wikipedia.org/wiki/John_Knoll

https://webfoundation.org/about/sir-tim-berners-lee/

https://www.biography.com/people/marc-andreessen-9542208

https://www.biography.com/people/jeff-bezos-9542209

https://en.wikipedia.org/wiki/PayPal_Mafia

http://www.notablebiographies.com/news/Ow-Sh/Page-Larry-and-Brin-Sergey.html#ixzz5FqodfQkZ

https://www.biography.com/people/mark-zuckerberg-507402

Patrick Aievoli

Patrick started his career in 1978 as a designer for local editorial and advertising companies. In 1984, he became a promotional designer at McGraw-Hill. During his time at McGraw-Hill Patrick helped in the creation of McGraw-Hill's first interactive CD-ROM "Encyclopedia of Science and Technology" in 1987. Professor Aievoli has been a full-time academic since 1989 when he left his position as senior designer - print promotion, at the McGraw-Hill Book Company.

In 1989 Prof. Aievoli started teaching full-time at SUNY Farmingdale immediately starting courses in multimedia. From 1990 to 1996 Prof. Aievoli completed his thesis on "The Use of New Media in Higher Education" culminating in an interactive art history CD-ROM featuring core and dynamic content along with a simplified suite of online learning tools. In 1998 he became a full-time faculty member at LIU Post in Brookville, NY

He became the director of the campus' Interactive Multimedia Arts graduate program in 1999 and built the program from its start to it's closing in 2018.

For the last 25 years Prof. Aievoli has helped several start-ups launch their interactive portfolios and has been in the online space developing websites, platforms, interactive television products, and apps/games.

Although he is a dedicated academic, Professor Aievoli is still involved in the new media arena and has consulted for some of the metro area's largest new media companies.

Client Roster (Partial)
Directly involved in the conception, creation and final production of numerous new media projects for companies:
- American Express
- Autism Academy
- FleetBoston
- McGraw-Hill Health Professions
- New York Islanders
- Verizon (NYNEX)
- TimeWarner/SONY
- Tommy Hilfiger USA

Recent Publications:
"Veal – breeding a new consumer class"
2016, Zeabook
University of Nebraska at Lincoln
Scholarly Library, Lincoln, Nebraska

"The Digital Incunabula: rock • paper • pixels"
2015, Zeabook
University of Nebraska at Lincoln
Scholarly Library, Lincoln, Nebraska

"on enterFrame"
2008, Whittier Publications
Long Beach, NY

"Supporting the Aesthetic through Metaphorical Thinking",
2004, Journal of National Collegiate Honors Council

"Colliding Forces", Chapter "Collide",
2004, McNabb and Kremer, Kendall Hunt